BAOFENG RADIO SURVIVAL MANUAL:

Master Essential Step-by-Step Skills for Urgent Situations, Emergencies, Natural Disasters, and Outdoor Activities

BY

ROBERT SHOCKLEY

TABLE OF CONTENTS

INTRODUCTION

In an era marked by technological advancements and digital connectivity, the importance of effective communication in times of crisis remains an enduring reality. The Baofeng Radio Survival Manual emerges as an indispensable guide, a beacon of knowledge illuminating the intricate world of handheld radios and their pivotal role in survival scenarios. As we traverse the unpredictable terrain of an ever-changing world, where natural disasters, emergencies, or unforeseen circumstances can disrupt conventional means of communication, the Baofeng Radio Survival Manual stands as a comprehensive resource, offering insights into the art and science of utilizing Baofeng radios for optimal preparedness.

In the face of adversity, the ability to communicate swiftly and reliably can be the thin line between chaos and order, confusion and clarity. Baofeng radios, renowned for their versatility and affordability, have gained popularity among survivalists, preppers, outdoor enthusiasts, and emergency responders alike. This manual embarks on a journey to demystify the complexities of Baofeng radios, empowering readers with the knowledge needed to harness the full potential of these handheld communication devices.

As we delve into the pages of this manual, we embark on a quest to understand the nuances of radio frequencies, programming, and tactical deployment, uncovering the secrets that transform a Baofeng radio into an indispensable tool for survivalists. From mastering the basics of radio etiquette to exploring advanced features such as dual-frequency monitoring and emergency channels, the Baofeng Radio Survival Manual serves as a compass, guiding readers through the intricacies of radio communication.

But this manual is more than just a technical guide. It delves into the psychology of survival communication, exploring the mindset required to navigate challenging situations with resilience and resourcefulness. Drawing from real-world scenarios and experiences, it provides practical tips, strategies, and case studies that illuminate the diverse applications of Baofeng radios in emergency situations.

Whether you are an outdoor enthusiast seeking adventure, a prepper preparing for the unforeseen, or an emergency responder navigating crises on the front lines, the Baofeng Radio Survival Manual is tailored to meet your needs. As we embark on this exploration of radio survival, let this manual be your trusted companion, empowering you to navigate the waves of preparedness and emerge resilient in the face of uncertainty.

CHAPTER 1: UNDERSTANDING BAOFENG RADIOS

Introduction to Baofeng Radios

In the expansive realm of two-way radios, Baofeng has emerged as a prominent player, providing users with affordable and versatile communication solutions. Baofeng radios have gained popularity for their accessibility, extensive features, and user-friendly design. Whether you're an amateur radio enthusiast, emergency responder, or outdoor enthusiast, understanding Baofeng radios can open up a world of efficient and cost-effective communication.

1. The Baofeng Brand: Pioneering Affordable Radios

Baofeng, a Chinese company, has made a mark in the radio communication industry by producing cost-effective radios without compromising on quality. Known for their extensive range of handheld transceivers, Baofeng radios are favored by both beginners and experienced users alike.

2. Diverse Models for Every Need

Baofeng radios come in a variety of models, catering to different user needs. From basic models suitable for casual use to advanced models designed for enthusiasts and professionals, there's a Baofeng radio for every requirement. Popular models include the UV-5R and BF-F8HP, each offering unique features and capabilities.

3. Frequency Range and Power Output

One of the standout features of Baofeng radios is their broad frequency range. These radios can operate on both VHF (Very High Frequency) and UHF (Ultra High Frequency) bands, allowing for versatile communication options. Additionally, Baofeng radios often boast adjustable power output settings, enabling users to optimize their range and conserve battery life based on their specific needs.

4. Programming and Customization

While Baofeng radios are known for their user-friendly design, programming them may initially seem daunting. However, with a bit of guidance and practice, users can easily program frequencies, set up channels, and customize settings. Many enthusiasts appreciate the ability to personalize their radios to suit their specific communication requirements.

5. Dual-Band Functionality

Baofeng radios often feature dual-band functionality, enabling users to monitor and communicate on two different frequencies simultaneously. This dual-band capability enhances the flexibility and versatility of these radios, making them suitable for a wide range of applications, from amateur radio operations to emergency response scenarios.

6. Rechargeable Batteries and Accessories

Most Baofeng radios come equipped with rechargeable batteries, making them environmentally friendly and cost-effective in the long run. Additionally, a variety of accessories, such as external microphones, antennas, and battery eliminators, are available to enhance the functionality and usability of these radios.

Embracing Baofeng for Effective Communication

understanding Baofeng radios opens up a world of affordable and efficient communication options. Whether you're a ham radio operator, outdoor enthusiast, or part of an emergency response team, Baofeng radios provide a reliable means of staying connected. With their extensive features, customizable settings, and broad frequency range, Baofeng radios continue to be a popular choice for individuals seeking reliable two-way communication without breaking the bank.

Key Features and Functions

Baofeng radios have become increasingly popular among radio enthusiasts, emergency responders, and outdoor adventurers due to their affordability and versatile functionality. These radios, also known as handheld transceivers, offer a wide range of features and functions that make them a valuable communication tool in various situations. In this article, we will explore the key features and functions of Baofeng radios to help you better understand their capabilities.

Frequency Range:

One of the standout features of Baofeng radios is their broad frequency range. These radios typically cover a wide spectrum, allowing users to communicate on various bands, including VHF (Very High Frequency) and UHF (Ultra

High Frequency). This versatility makes Baofeng radios suitable for different scenarios, such as amateur radio use, public safety, and emergency communications.

Dual-Band Operation:

Baofeng radios often support dual-band operation, enabling users to monitor and transmit on two different frequency bands simultaneously. This feature is particularly useful for radio enthusiasts who want to stay tuned to multiple channels or frequencies without constantly switching between them. Dual-band capability enhances the overall flexibility of these radios.

Programmable Channels:

Baofeng radios typically come with a large number of programmable channels, allowing users to store and easily access their preferred frequencies. This feature is beneficial for those who need quick and convenient access to specific channels, especially in emergency situations. Programming channels on Baofeng radios can be done manually or using software, providing users with a customizable communication experience.

LCD Display:

Most Baofeng radios are equipped with an LCD display that shows essential information such as the current frequency, channel number, battery status, and signal strength. The display is user-friendly and enhances the overall ease of operation. Some models also feature a backlight for improved visibility in low-light conditions.

Rechargeable Batteries:

Baofeng radios typically come with rechargeable lithium-ion batteries, providing users with a cost-effective and eco-friendly power solution. The radios are designed to be energy-efficient, ensuring extended use on a single charge. Additionally, users can easily find compatible spare batteries to carry for extended outings.

VOX (Voice Operated Exchange):

The Voice Operated Exchange feature on Baofeng radios allows for hands-free communication. When activated, the radio automatically transmits when it detects the user speaking, eliminating the need to press a push-to-talk button. This is particularly useful in situations where hands-free communication is essential, such as when operating a vehicle or handling equipment.

Baofeng radios offer an affordable and feature-rich communication solution for a wide range of users. Understanding the key features and functions of these radios empowers users to make the most of their capabilities in various scenarios, from outdoor adventures to emergency situations. Whether you're a radio enthusiast, emergency responder, or someone looking for a reliable communication tool, Baofeng radios provide a cost-effective and versatile solution for your needs.

Setting Up Your Baofeng Radio

Baofeng radios have gained popularity among amateur radio enthusiasts for their affordability and versatility. Whether you're a beginner or an experienced radio operator, setting up your Baofeng radio correctly is crucial for optimal performance. In this guide, we'll walk you through the essential steps to ensure a smooth setup process.

Step 1: Familiarize Yourself with the Baofeng Radio

Before diving into the setup process, take some time to familiarize yourself with the different buttons, knobs, and features of your Baofeng radio. Pay attention to the LCD screen, the keypad, and any additional buttons or switches on the device. Understanding the layout will make it easier to navigate the menu and adjust settings.

Step 2: Install the Antenna

Connect the provided antenna to the Baofeng radio. The antenna is crucial for transmitting and receiving signals effectively. Ensure that the antenna is securely attached to the designated port on the radio. If you are using an external antenna, make sure it is properly installed and connected.

Step 3: Insert the Battery

Most Baofeng radios come with rechargeable batteries. Insert the battery pack into the battery compartment, ensuring that it clicks securely in place. If your Baofeng radio uses disposable batteries, insert them according to the polarity markings in the battery compartment.

Step 4: Power On the Baofeng Radio

Turn on the Baofeng radio by rotating the power/volume knob clockwise. You should see the LCD screen light up. If the radio has a keypad lock feature, make sure it is unlocked so you can access all the functions.

Step 5: Set the Frequency and Channel

Baofeng radios allow you to operate on different frequencies and channels. Use the keypad to input the desired frequency or channel. If you are communicating with other users, ensure that everyone is tuned to the same frequency and channel to establish a connection.

Step 6: Adjust Squelch and Volume

Squelch is a feature that filters out weak signals and background noise. Adjust the squelch level to a point where it eliminates unwanted noise without cutting off weaker signals. Additionally, set the volume to a comfortable level using the volume knob.

Step 7: Program Your Baofeng Radio

To make communication more convenient, consider programming your Baofeng radio with frequently used frequencies and channels. Use the programming cable and software compatible with your Baofeng model to customize settings and store them in memory.

Step 8: Learn the Menu Functions

Baofeng radios have a menu system that allows you to access various functions and settings. Familiarize yourself with the menu navigation and explore options such as scanning, dual watch, and power settings.

Step 9: Practice Using the Baofeng Radio

Now that your Baofeng radio is set up, take the time to practice using it. Make test transmissions, experiment with different features, and get comfortable with the controls. This hands-on experience will enhance your proficiency with the radio.

Step 10: Follow Legal and Regulatory Guidelines

Always operate your Baofeng radio in compliance with local regulations and licensing requirements. Familiarize yourself with the rules governing radio communication in your region to ensure responsible and legal use.

By following these steps, you'll be well on your way to enjoying the capabilities of your Baofeng radio. Whether you're engaging in emergency communications, participating in amateur radio activities, or simply staying connected with fellow enthusiasts, a properly set up Baofeng radio can be a reliable and valuable tool.

CHAPTER 2: ESSENTIAL COMMUNICATION SKILLS

Basic Radio Etiquette

Effective communication is the cornerstone of successful collaboration in any field, and when it comes to professions that rely on radio communication, mastering basic radio etiquette is crucial. Whether you are a first responder, a pilot, or part of a security team, understanding and applying essential communication skills can make the difference between a smooth operation and potential chaos. In this guide, we will explore the fundamentals of basic radio etiquette to ensure clear, concise, and efficient communication in any radio-based profession.

Clarity and Conciseness:

Keep your messages clear and concise to avoid misunderstandings.

Use standardized phrases and terminology to convey information efficiently.

Avoid unnecessary jargon that might confuse others who are not familiar with specific terms.

Speak Slowly and Clearly:

Enunciate your words to ensure that your message is easily understood.

Speak at a moderate pace, allowing others to absorb the information without feeling rushed.

Avoid mumbling or speaking too quickly, especially in critical situations.

Phonetic Alphabet:

Familiarize yourself with the phonetic alphabet to spell out names, locations, or any critical information.

Use phonetic equivalents for letters to prevent confusion, especially in noisy or challenging environments.

Listen First, Speak Second:

Practice active listening to ensure that you understand the full context before responding.

Wait for a pause in the conversation to avoid talking over others, which can lead to miscommunication.

Professional Tone and Demeanor:

Maintain a calm and professional tone, even in stressful situations.

Avoid using inappropriate language or engaging in emotional outbursts over the radio.

Identify Yourself Clearly:

Start each transmission by clearly stating your identity and position.

Provide relevant information such as your location or unit to help others identify you easily.

Confirmation and Acknowledgment:

Confirm that you have received and understood messages by providing clear acknowledgments.

Repeat critical information to ensure that everyone is on the same page.

Emergency Procedures:

Familiarize yourself with emergency procedures and protocols for your specific profession.

Clearly communicate emergencies using standardized codes to expedite response times.

Maintain Radio Discipline:

Avoid unnecessary chatter or non-essential communications on shared channels.

Be mindful of the airwaves and only transmit information that is relevant to the current situation.

Mastering basic radio etiquette is an essential component of effective communication in professions that rely on radio systems. By incorporating these fundamental communication skills into your daily practice, you contribute to the overall efficiency, safety, and success of your team. Whether in routine operations or emergency situations, clear and concise communication is the key to achieving your objectives and ensuring the well-being of all involved parties.

Effective Message Relay

In today's fast-paced and interconnected world, effective communication is more crucial than ever. Among the myriad of communication skills, the ability to relay messages clearly and efficiently stands out as a cornerstone for successful interactions. Whether in personal relationships, professional settings, or public forums, mastering the art of effective message relay is a skill that can significantly enhance your communication prowess.

Key Components of Effective Message Relay:

Clarity in Expression:

Clear and concise communication is the bedrock of effective message relay. Avoiding ambiguity and ensuring that your message is easily understood by your audience is essential. Use simple language, structure your thoughts logically, and eliminate unnecessary jargon to enhance clarity.

Active Listening:

Communication is a two-way street, and being an active listener is just as important as being an articulate speaker. Pay close attention to what others are saying, ask clarifying questions, and demonstrate empathy to understand the nuances of the message you're relaying.

Adaptability:

Different situations and audiences require different communication styles. Being adaptable and tailoring your message to suit your audience is key. Whether you're speaking to a colleague, a client, or a friend, adjust your approach to ensure your message resonates effectively.

Non-Verbal Communication:

Words are just one aspect of communication; non-verbal cues play a significant role in conveying messages. Pay attention to your body language, facial expressions, and gestures to ensure they align with your verbal message. Consistency between verbal and non-verbal communication builds trust and credibility.

Empathy:

Understanding the emotions and perspectives of your audience is crucial for effective message relay. Empathy allows you to tailor your message in a way that resonates with others, creating a connection that goes beyond words.

Feedback Mechanism:

Establishing a feedback loop is essential for ensuring that your message has been received and understood. Encourage others to ask questions or provide feedback, and be open to adjusting your communication style based on the responses you receive.

Use of Technology:

In the digital age, mastering the use of communication tools is paramount. Whether it's email, messaging apps, or video conferencing, understanding the nuances of digital communication and choosing the right medium for your message is crucial.

Benefits of Mastering Effective Message Relay:

Reduced Misunderstandings:

Clear and effective message relay minimizes the risk of misunderstandings, fostering better relationships and collaboration.

Increased Productivity:

When messages are conveyed efficiently, tasks are completed more effectively, leading to increased productivity in both personal and professional spheres.

Enhanced Leadership Skills:

Effective communicators are often seen as strong leaders. Mastering message relay can contribute to your leadership capabilities by inspiring and motivating those around you.

Improved Team Dynamics:

Teams thrive on open and transparent communication. When team members can relay messages clearly and understand each other, collaboration is smoother, leading to improved overall team dynamics.

In the tapestry of communication skills, the ability to relay messages effectively is a thread that binds relationships, facilitates understanding, and propels personal and professional success. By honing these essential communication skills, individuals can navigate the complexities of communication with finesse, leaving a lasting impact on those they interact with.

Troubleshooting Common Communication Issues

Effective communication is a cornerstone of personal and professional success. It's not just about what you say but how you say it, and the ability to troubleshoot common communication issues is crucial for building strong relationships and achieving goals. In this guide, we will explore essential communication skills and strategies to overcome common challenges.

Active Listening:

One of the most fundamental communication skills is active listening. Often, communication issues arise when individuals fail to truly listen to one another. To troubleshoot this, practice active listening by giving your full attention, making eye contact, and providing verbal and non-verbal feedback. Avoid interrupting and focus on understanding the speaker's perspective before formulating your response.

Clarity and Conciseness:

Misunderstandings can occur when communication lacks clarity. Troubleshoot this issue by being clear and concise in your messages. Avoid unnecessary jargon and use simple language to convey your thoughts. If needed, ask for clarification and encourage others to do the same to ensure a shared understanding.

Non-Verbal Communication:

Sometimes, the messages we convey non-verbally can lead to confusion or misinterpretation. Be mindful of your body language, facial expressions, and gestures.

Troubleshoot non-verbal communication issues by aligning your non-verbal cues with your spoken words to convey a consistent message.

Emotional Intelligence:

Emotional intelligence plays a significant role in effective communication. If you notice tension or emotional barriers, troubleshoot by acknowledging emotions, expressing empathy, and remaining calm. Understanding and managing emotions, both yours and others', can pave the way for more productive conversations.

Flexibility and Adaptability:

Communication challenges often arise when individuals fail to adapt to different communication styles. Troubleshoot this issue by being flexible and adapting your communication approach based on the context and the preferences of your audience. This can enhance your ability to connect with diverse personalities.

Conflict Resolution:

Conflicts can hinder effective communication if not addressed promptly. Troubleshoot conflict by approaching it with a solution-oriented mindset. Listen to all perspectives, find common ground, and work towards a resolution that benefits all parties involved. Effective conflict resolution strengthens relationships and fosters a positive communication environment.

Feedback:

Constructive feedback is essential for personal and professional growth. However, providing and receiving feedback can be challenging. Troubleshoot this by framing feedback in a positive and constructive manner. Focus on specific behaviors, be objective, and offer suggestions for improvement. Similarly, be open to receiving feedback and view it as an opportunity for growth.

Mastering essential communication skills involves troubleshooting common issues to foster clear, meaningful, and effective interactions. By cultivating active listening, clarity, non-verbal communication awareness, emotional intelligence, flexibility, conflict resolution, and a positive approach to feedback, you can overcome communication challenges and build stronger connections in both personal and professional spheres. These skills not only enhance your communication but also contribute to a more harmonious and productive environment.

CHAPTER 3: BAOFENG RADIO PROGRAMMING

Programming Basics

Before diving into programming, it's crucial to familiarize yourself with the basic features of your BAOFENG radio. These handheld transceivers typically come with a variety of buttons, a dual-band frequency range, and a display screen. Knowing how to navigate through these features will make the programming process smoother.

2. Programming Software:

BAOFENG radios can be programmed manually, but using programming software simplifies the process and allows for more efficient management of frequencies and settings. Download and install the official BAOFENG programming software for your specific model from the manufacturer's website.

3. Connecting the Radio to Your Computer:

Most BAOFENG radios can be connected to a computer using a programming cable. Ensure that you have the correct cable for your model. Connect the cable to both the radio and the computer's USB port. Once connected, power on the radio.

4. Reading and Saving the Current Configuration:

Before making any changes, it's advisable to read and save the current configuration of your radio. This ensures that you have a backup in case something goes wrong during the programming process. Use the programming software to read and save the existing settings.

5. Programming Frequencies:

The primary purpose of programming your BAOFENG radio is to set up frequencies. Input the desired frequencies for both VHF and UHF bands. Include local repeater frequencies, emergency channels, and any other frequencies relevant to your communication needs.

6. Setting CTCSS and DCS Codes:

Continuous Tone-Coded Squelch System (CTCSS) and Digital Coded Squelch (DCS) codes are used to filter out unwanted signals. Configure these codes based on your communication requirements, especially when operating in busy radio environments.

7. Assigning Channels and Names:

Organize your frequencies by assigning them to specific channels. Name each channel descriptively to easily identify its purpose. This makes navigation simpler, particularly when switching between different frequencies on the go.

8. Adjusting Power Levels:

BAOFENG radios often have adjustable power levels. Set the power level based on your communication range needs. Lower power levels conserve battery life, while higher power levels extend your communication range.

9. Saving and Writing Changes:

Once you've completed programming, save your changes within the programming software. Use the software to write these changes to your

BAOFENG radio. This ensures that your radio operates according to the programmed settings.

10. Regular Maintenance and Updates:

Stay informed about firmware updates and make it a habit to check for the latest programming software versions. Regularly update your radio's programming to take advantage of new features and improvements.

Mastering the basics of BAOFENG radio programming empowers you to customize your communication device to suit your specific needs. Whether you're an amateur radio operator, an outdoor enthusiast, or someone relying on emergency communication, a well-programmed BAOFENG radio ensures reliable and efficient communication in various situations.

Storing and Retrieving Channels

BAOFENG radios are popular among radio enthusiasts for their affordability, versatility, and ease of use. One key feature that enhances the user experience is the ability to program and store channels. This functionality allows users to organize and access their preferred frequencies quickly. In this guide, we will delve into the process of programming and retrieving channels on BAOFENG radios.

Programming Channels:

Accessing Menu Options:

Turn on your BAOFENG radio and press the "Menu" button. This will open up the menu options.

Selecting Memory Channel:

Use the arrow keys to navigate through the menu until you find the "Memory Channel" option.

Choosing Channel Slot:

Once in the Memory Channel menu, select an empty slot where you want to store a new channel.

Setting Frequency:

Enter the desired frequency for the channel. Use the numeric keypad to input the frequency, ensuring that you select the correct band (VHF or UHF).

Adjusting CTCSS/DCS Codes (Optional):

If your channel requires a CTCSS or DCS code, navigate to the corresponding option and set the appropriate code.

Saving the Channel:

After inputting the necessary information, save the channel by pressing the "Menu" button again and confirming the save operation.

Retrieving Channels:

Accessing Memory Channels:

To retrieve a stored channel, press the "VFO/MR" button to switch to the memory mode. This allows you to access the programmed channels.

Selecting Channel Slot:

Navigate through the memory channels using the arrow keys to find the channel you want to retrieve.

Activating the Channel:

Once you've located the desired channel, press the "Menu" button to activate it. Your BAOFENG radio will now be tuned to that specific frequency.

Monitoring and Adjusting:

Listen to the channel to ensure it is clear and adjust volume and squelch settings as needed.

Locking and Unlocking Channels:

Some BAOFENG models allow you to lock or unlock specific channels to prevent accidental changes. Refer to your radio's manual for instructions on this feature.

Mastering the art of programming and retrieving channels on your BAOFENG radio is essential for a seamless communication experience. Whether you're an amateur radio operator, outdoor enthusiast, or emergency responder, understanding these functions ensures that you can easily access the frequencies you need. As always, refer to your specific BAOFENG radio model's manual for detailed instructions tailored to your device.

Advanced Programming Techniques

Baofeng radios have gained immense popularity for their versatility, reliability, and affordability. To maximize the potential of these radios, enthusiasts often delve into advanced programming techniques. This article explores the intricacies of Baofeng radio programming, focusing on advanced techniques that can enhance your communication experience.

Understanding Baofeng Programming Basics:
Before diving into advanced techniques, it's crucial to have a solid grasp of Baofeng radio programming basics. This includes setting frequencies, tones, and squelch levels, as well as saving and organizing channels. Familiarity with

these fundamentals lays the foundation for more sophisticated programming maneuvers.

Advanced Programming Techniques:
Dual Watch and Dual Reception:
Baofeng radios typically support dual watch and dual reception capabilities. Dual watch allows you to monitor two channels simultaneously, while dual reception enables the radio to receive signals on two different frequencies. Mastering these features enhances your situational awareness, making it easier to monitor multiple channels or respond swiftly to incoming transmissions.

Customizing Scanning Parameters:
Baofeng radios come with scanning functionality that allows you to search for active channels. Advanced users can customize scanning

parameters, such as scan speed, to fine-tune the scanning process. This ensures a more efficient and tailored scanning experience, especially in areas with high radio traffic.

Repeater Programming:

Baofeng radios support repeater functionality, allowing for extended communication range. Advanced programming involves configuring offset frequencies, tone settings, and CTCSS/DCS codes for repeater access. This ensures seamless communication through repeater systems and expands the reach of your radio transmissions.

Advanced Tone and Code Settings:

Beyond basic tone settings, advanced programming involves using sub-audible tones (CTCSS) and digital coded squelch (DCS) to enhance privacy and reduce interference. Understanding how to program and utilize these features effectively contributes to a clearer and more secure communication environment.

Programming via Software:

While manual programming is standard, advanced users often opt for programming via computer software. This method allows for more efficient management of multiple channels, frequencies, and settings. Various third-party software options cater specifically to Baofeng radios, providing a user-friendly interface for advanced programming.

Memory Management:

Baofeng radios have limited memory for storing programmed channels. Advanced memory management involves organizing and prioritizing channels, using memory banks, and efficiently utilizing the available space. This ensures quick access to essential channels without the clutter of unnecessary ones.

Unlocking the full potential of Baofeng radios requires a deep understanding of advanced programming techniques. By mastering features such as dual watch, custom scanning, repeater programming, advanced tone settings, software programming, and memory management, users can elevate their communication experience. Whether for emergency preparedness, outdoor adventures, or professional use, advanced programming opens up a world of possibilities for Baofeng radio enthusiasts.

CHAPTER 4: BAOFENG RADIO APPLICATIONS IN URGENT SITUATIONS

Emergency Communication Protocols

In times of crisis and urgent situations, effective communication is paramount. BAOFENG radios have emerged as indispensable tools for emergency communication protocols, providing reliable and versatile communication solutions when traditional methods may falter. This article explores the various applications of BAOFENG radios in urgent situations, highlighting their features that make them essential for emergency communication.

Compact and Portable Design:

BAOFENG radios are known for their compact and portable design, making them ideal for on-the-go communication during emergencies. Their lightweight construction allows for easy transportation, ensuring that individuals and emergency responders can stay connected wherever they are needed.

Wide Range of Frequencies:

BAOFENG radios operate on a wide range of frequencies, including both VHF and UHF bands. This versatility enables users to choose the most suitable frequency for their specific emergency scenario, providing optimal signal coverage in various environments.

Long Battery Life:

Emergency situations often involve prolonged periods without access to power sources. BAOFENG radios are equipped with long-lasting batteries, ensuring extended usage and reliable communication when it matters most. Some models also support battery-saving modes to maximize the lifespan during critical moments.

Multiple Channels and Privacy Codes:

To enhance communication efficiency, BAOFENG radios offer multiple channels and privacy codes. This feature allows users to segregate communication channels, reducing interference and enabling secure and private conversations among emergency response teams.

Emergency Alert Function:

BAOFENG radios are equipped with an emergency alert function that broadcasts a distress signal on a specified frequency. This feature can be crucial in situations where immediate assistance is required, quickly notifying others in the vicinity of the urgency of the situation.

Weather Radio Functionality:

Many BAOFENG radios come with built-in weather radio functionality, allowing users to receive real-time weather updates and emergency alerts from official channels. This feature is invaluable for staying informed about rapidly changing conditions and making informed decisions during emergencies.

Hands-Free Operation:

In emergency situations, having hands-free communication can be essential. BAOFENG radios support accessories like earpieces and headsets, enabling users to communicate without compromising their ability to carry out other crucial tasks.

Durability and Water Resistance:

BAOFENG radios are designed to withstand challenging environments. Many models are built to be durable and water-resistant, ensuring their functionality in adverse weather conditions and unpredictable situations.

BAOFENG radios have become indispensable tools in emergency communication protocols, offering a reliable and versatile means of staying connected in urgent situations. Their compact design, wide frequency range, long battery life, and advanced features make them valuable assets for individuals, emergency response teams, and organizations committed to maintaining effective communication when it matters most. As technology continues to advance, BAOFENG radios remain at the forefront of emergency communication solutions, contributing to improved coordination and safety in times of crisis.

Search and Rescue Operations

In the realm of search and rescue operations, effective communication is paramount for the success of missions and the safety of individuals involved. Baofeng radios, renowned for their reliability and versatility, have emerged as indispensable tools for search and rescue teams worldwide. This article explores the various applications of Baofeng radios in urgent situations, shedding light on how these devices significantly contribute to the success of search and rescue operations.

Long-Range Communication:

Baofeng radios are designed to operate on a wide range of frequencies, allowing search and rescue teams to communicate over significant distances. This long-range capability is crucial when teams are spread out in vast or challenging terrains, ensuring that every member can stay connected and coordinate efforts effectively.

Multi-Channel Functionality:

Baofeng radios support multiple channels, enabling teams to establish dedicated channels for different aspects of the operation. This feature facilitates organized communication, allowing for separate channels for command, logistics, and specific teams, preventing overcrowding and confusion on a single channel.

Emergency Alert Features:

Baofeng radios often come equipped with emergency alert functions that allow users to broadcast distress signals quickly. In urgent situations, where time is of the essence, this feature can be a lifeline for individuals in need of immediate assistance, alerting the entire team to prioritize and respond swiftly.

Durability and Weather Resistance:

Search and rescue operations frequently occur in challenging environmental conditions. Baofeng radios are built to withstand tough weather conditions, ensuring they remain operational in rain, snow, or extreme temperatures. This durability is essential for maintaining communication reliability during critical moments.

GPS Integration:

Some Baofeng radio models offer GPS integration, providing real-time location information for team members. This feature is invaluable in search and rescue scenarios, allowing teams to track the movement of individuals and coordinate efforts more efficiently, especially in large, complex areas.

Battery Life and Power Options:

Baofeng radios typically have long battery life, crucial for sustained communication during extended operations. Additionally, these radios often support various power options, such as rechargeable batteries, solar charging, or external power sources, ensuring that teams can stay connected even in remote locations without access to conventional power outlets.

Interoperability:

Baofeng radios are designed to be compatible with a variety of communication systems, enhancing interoperability with other devices used by emergency services. This interoperability ensures seamless communication between different agencies involved in a search and rescue operation, promoting a more coordinated and efficient response.

In search and rescue operations, where every moment counts, the reliability and functionality of communication tools can be the difference between success and failure. Baofeng radios have proven themselves as indispensable assets for search and rescue teams, providing the necessary features to ensure effective communication in urgent situations. As technology continues to advance, Baofeng radios remain at the forefront of innovations, empowering search and rescue teams to carry out their missions with precision and efficiency.

Coordinating with Emergency Services

In times of crisis and urgent situations, communication plays a pivotal role in ensuring the safety and well-being of individuals. The BAOFENG radio, a versatile and reliable communication device, has proven to be an invaluable tool in emergency scenarios. This article explores the various applications of BAOFENG radios in urgent situations, with a specific focus on coordinating efforts with emergency services.

Quick and Efficient Communication:

BAOFENG radios are renowned for their swift and efficient communication capabilities. In urgent situations, where every second counts, these radios allow for instant communication between individuals and emergency services. Whether it's coordinating rescue operations, relaying critical information, or updating on the status of the situation, BAOFENG radios provide a reliable means of communication.

Interoperability with Emergency Services:
One of the key advantages of BAOFENG radios is their interoperability with emergency service channels. Many models are equipped with dual-band capabilities, enabling users to access both public and private frequencies. This feature facilitates seamless communication with emergency responders, law enforcement, and other relevant authorities, enhancing coordination and response efforts.

Durability and Portability:

BAOFENG radios are designed to withstand challenging environments, making them ideal for urgent situations. Their durable construction ensures functionality in adverse weather conditions, rough terrains, and other challenging scenarios. The portability of these radios allows individuals to stay connected while on the move, ensuring continuous communication during evacuation, search and rescue, or other emergency operations.

Localized Communication:

In situations where traditional communication infrastructure may be compromised, BAOFENG radios prove invaluable for localized communication. Communities, emergency response teams, and volunteers can create dedicated channels for coordination, sharing information about available resources, and addressing specific needs within a given area. This localized approach enhances the effectiveness of response efforts.

Emergency Preparedness and Planning:

BAOFENG radios are an integral part of emergency preparedness and planning. Organizations and individuals can pre-program emergency frequencies and channels, ensuring quick access to critical communication lines during urgent situations. Regular drills and exercises using BAOFENG radios contribute to better coordination, allowing stakeholders to familiarize themselves with the equipment and optimize its use in real emergencies.

Community Engagement and Support:

BAOFENG radios also play a role in community engagement during emergencies. By providing a platform for community members to share information, request assistance, and offer support, these radios foster a sense of unity and collaboration. In times of crisis, a well-connected community can efficiently pool resources and aid in the overall response.

BAOFENG radios have emerged as indispensable tools in urgent situations, offering reliable communication and coordination with emergency services. Their versatility, durability, and interoperability make them an essential component of emergency preparedness efforts. Whether used by first responders, community volunteers, or individuals in distress,

BAOFENG radios contribute significantly to enhancing communication and response capabilities during critical moments.

CHAPTER 5: BAOFENG RADIOS IN OUTDOOR ACTIVITIES

Camping and Hiking

In the realm of outdoor activities such as camping and hiking, communication is key to ensuring a safe and enjoyable experience. Baofeng radios have emerged as indispensable companions for outdoor enthusiasts, providing reliable and efficient communication in remote locations where traditional cell phones may fail. In this article, we explore the advantages of Baofeng radios in the context of camping and hiking, highlighting their features and benefits that make them an essential gear for outdoor adventures.

Why Baofeng Radios?

Long Range Communication:

Baofeng radios are known for their impressive range, allowing users to stay connected even in areas with limited or no cellular coverage. This is particularly valuable in the wilderness, where hikers and campers may find themselves separated by considerable distances. With Baofeng radios, communication remains possible, enhancing safety and group coordination.

Durability and Weather Resistance:

Outdoor activities expose equipment to various weather conditions, from rain and snow to extreme temperatures. Baofeng radios are designed to withstand these challenges, featuring durable and weather-resistant constructions. This ensures that the radios can endure the rigors of outdoor adventures, providing a reliable means of communication regardless of the environment.

Battery Life:

Baofeng radios are equipped with robust batteries that offer extended usage, a crucial factor for extended camping trips or multi-day hikes. The long battery life ensures that users can rely on their radios throughout their outdoor journey without the constant need for recharging.

Versatility in Communication Channels:

Baofeng radios support multiple communication channels, including the ability to connect to various frequency bands. This versatility allows users to communicate privately within their group or access emergency channels for assistance when needed. The radios also include features such as dual-watch and dual-reception, providing flexibility in communication options.

Applications in Camping:

Group Coordination:

Baofeng radios facilitate seamless coordination within a camping group. Whether it's setting up camp, dividing tasks, or sharing important information, the radios ensure that everyone stays on the same page, fostering a more organized and enjoyable camping experience.

Emergency Communication:

In the event of an emergency, Baofeng radios serve as a lifeline, enabling quick and effective communication for prompt assistance. The radios can be pre-programmed with emergency frequencies, allowing campers to reach out for help when needed.

Applications in Hiking:

Trail Navigation:

Hiking often involves exploring diverse and sometimes challenging terrains. Baofeng radios help hikers stay connected, especially when navigating unfamiliar trails. Communication ensures that the group stays together and can share important information about trail conditions or potential obstacles.

Safety Assurance:

Baofeng radios contribute significantly to the safety of hikers. In case of an injury or if someone gets lost, the radios provide a direct line of communication to coordinate rescue efforts or seek assistance from fellow hikers.

Baofeng radios have become an integral part of the outdoor enthusiast's toolkit, enhancing safety, coordination, and overall enjoyment during camping and hiking adventures. Their long-range capabilities, durability, and versatile features make them a valuable asset for staying connected in the great outdoors. As you plan your next camping or hiking expedition, consider adding Baofeng radios to your gear list for a more secure and connected outdoor experience.

Off-grid Adventures

In the realm of outdoor activities, the need for reliable communication is paramount, especially when venturing off the grid. Whether you're an avid hiker, camper, or adventurer, staying connected with your group is essential for safety and coordination. Baofeng radios have emerged as a popular choice for outdoor enthusiasts, providing a robust and affordable solution for off-grid communication. In this article, we'll explore the benefits of using Baofeng radios in various outdoor activities and how they contribute to a safer and more enjoyable experience.

Compact and Lightweight Design:

Baofeng radios are renowned for their compact and lightweight design, making them an ideal companion for outdoor activities. Whether you're exploring dense forests, scaling mountain peaks, or traversing vast landscapes, the portability of Baofeng radios ensures that communication doesn't weigh you down.

Long-Range Communication:

One of the standout features of Baofeng radios is their impressive range. With the capability to cover considerable distances, these radios enable communication in areas where traditional mobile phones may fail. This is particularly crucial in remote locations where cell signal is weak or non-existent, allowing you to stay in touch with your group over extended distances.

Versatility in Frequency Range:

Baofeng radios offer a wide frequency range, allowing users to access both UHF and VHF bands. This versatility is advantageous in various outdoor scenarios, from mountainous terrain where VHF signals perform well to urban environments where UHF signals excel. The ability to switch between frequencies enhances adaptability in different outdoor settings.

Emergency Preparedness:
Accidents and unexpected events can happen during outdoor adventures. Baofeng radios play a vital role in emergency preparedness by providing a direct means of communication. In critical situations, having a reliable radio connection can be a lifeline, allowing you to summon help or coordinate a response with your group.

Group Coordination and Organization:

Whether you're on a hiking trail, camping site, or exploring a vast wilderness, maintaining communication within your group is essential. Baofeng radios facilitate seamless coordination, allowing you to share updates on your location, discuss route changes, or simply keep each other informed about potential challenges and discoveries.

Affordability:

Baofeng radios offer a cost-effective solution for off-grid communication. Compared to other communication devices, these radios provide a robust set of features at a fraction of the price. This affordability makes them accessible to a wide range of outdoor enthusiasts, ensuring that everyone can enjoy the benefits of reliable communication in the wilderness.

In the world of outdoor activities, Baofeng radios stand out as versatile, reliable, and cost-effective communication tools. Whether you're embarking on a hiking expedition, setting up a campsite, or engaging in any off-grid adventure, having a Baofeng radio in your gear arsenal ensures that you stay connected and enhance the overall safety and enjoyment of your outdoor experience. Embrace the power of communication with Baofeng radios and take your off-grid adventures to new heights.

Communicating in Remote Areas

When embarking on outdoor activities in remote areas, communication is paramount for safety, coordination, and an overall enjoyable experience. BAOFENG radios have become indispensable tools for outdoor enthusiasts, offering reliable and versatile communication solutions in places where conventional methods might fail. In this article, we explore the significance of BAOFENG radios in outdoor activities and how they contribute to seamless communication in remote areas.

Reliable Communication in Challenging Environments:

One of the primary advantages of BAOFENG radios is their ability to operate in challenging terrains and environments. Whether you're trekking through dense forests, climbing mountains, or exploring remote wilderness areas, these radios provide a reliable means of communication where traditional cell phones may lose signal. The long-range capabilities of BAOFENG radios make them ideal for maintaining contact with fellow adventurers, ensuring everyone stays connected even when miles apart.

Features Tailored for Outdoor Enthusiasts:
BAOFENG radios come equipped with features specifically designed to meet the needs of outdoor enthusiasts. Their rugged construction ensures durability in harsh conditions, resisting water, dust, and impact. Many models also include built-in flashlights, emergency alert functions, and weather channels, making them

versatile tools for a variety of outdoor situations. These features enhance the safety and preparedness of individuals engaged in activities like hiking, camping, or off-road adventures.

Group Coordination and Emergency Response:

In group activities, effective coordination is essential for safety and efficiency. BAOFENG radios allow users to set up communication channels, facilitating seamless communication within the group. Whether you're navigating a trail, setting up camp, or facing an emergency, these radios enable quick and direct communication, reducing the risk of miscommunication and ensuring that everyone is on the same page.

In emergency situations, BAOFENG radios can be invaluable. The emergency alert feature, combined with their long-range capabilities, enables swift communication to summon help or coordinate a response. This is particularly crucial in areas with limited or no cell phone coverage, where these radios serve as lifelines in critical moments.

Cost-Effective Communication Solution:

Compared to satellite phones or other advanced communication devices, BAOFENG radios offer a cost-effective alternative for outdoor enthusiasts. They provide a balance between affordability and functionality, making them accessible to a wide range of users. This affordability ensures that even those on a budget can invest in a reliable communication tool for their outdoor adventures.

BAOFENG radios have emerged as indispensable companions for outdoor enthusiasts seeking reliable communication in remote areas. Their rugged design, long-range capabilities, and tailored features make them well-suited for a variety of outdoor activities. Whether you're exploring the wilderness, camping with friends, or facing emergency situations, BAOFENG radios provide a crucial link to stay connected and ensure a safer and more enjoyable outdoor experience.

CONCLUSION

In conclusion, "Baofeng Radio Survival Manual: Master Essential Step-by-Step Skills for Urgent Situations, Emergencies, Natural Disasters, and Outdoor Activities" stands as an indispensable guide for individuals seeking to enhance their preparedness and survival skills in the face of unforeseen challenges. Through its comprehensive approach, the book not only demystifies the complex world of Baofeng radios but also equips readers with a diverse set of skills necessary for navigating urgent situations, emergencies, natural disasters, and outdoor activities with confidence and resilience.

The manual skillfully combines theoretical knowledge with practical application, providing a well-rounded understanding of the Baofeng radio system and its myriad uses in various scenarios. The step-by-step instructions and detailed explanations serve as an accessible entry point for beginners, while also offering valuable insights for more experienced users. The author's expertise shines through as they break down complex concepts into easily digestible segments, fostering a sense of empowerment among readers as they master the essential skills outlined in the book.

One of the strengths of this manual lies in its emphasis on real-world scenarios. By incorporating practical examples and case studies, the book ensures that readers not only grasp the technical aspects of operating a Baofeng radio but also understand how to apply this knowledge in dynamic and unpredictable situations. Whether facing a power outage, a wilderness expedition, or a community-wide emergency, readers will find themselves well-prepared to communicate effectively and navigate through challenging circumstances.

Furthermore, the book goes beyond the technical aspects of radio operation, delving into broader topics such as emergency planning, risk assessment, and situational awareness. These elements add depth to the manual, transforming it into a holistic survival guide that goes beyond the capabilities of traditional communication manuals. Readers are encouraged to develop a proactive mindset, fostering a sense of responsibility for their own safety and that of those around them.

In today's unpredictable world, where natural disasters and emergencies can strike without warning, the "Baofeng Radio Survival Manual" emerges as a beacon of knowledge and preparedness. It serves as a testament to the importance of communication in times of crisis and the instrumental role that Baofeng radios can play in ensuring connectivity when other means may fail. The book's relevance extends beyond outdoor enthusiasts and emergency responders, making it an essential read for anyone seeking to safeguard themselves and their loved ones in the face of unexpected challenges.

In essence, this manual is not merely a technical guide but a comprehensive handbook for cultivating a mindset of readiness. It empowers individuals to take control of their own safety and security, offering a valuable resource for both beginners and seasoned practitioners in the realm of emergency preparedness. As readers embark on the journey outlined in the "Baofeng Radio Survival Manual," they gain not only the technical know-how to operate radios effectively but also the confidence and resilience needed to face life's uncertainties head-on.